Business Math

A learning workbook program for junior high and high school students.

ISBN-13: 978-1725514041
ISBN-10: 1725514044

Printed by CreateSpace, An Amazon.com Company.
Available from Amazon.com and other retail outlets.
CreateSpace, Charleston, SC

Consult a professional when seeking business advice and decisions. This is a learning book discussing
topics in a general style, not intended to be considered professional advice, suggestions. or guidance.
Content Updated: 2020
Submit all inquiries at the website www.YMBAgroup.com

Y.M.B.A. Business Math - grades 6 7 8 9 10 +   ages 12 13 14 15 16 +

Business Math

We hope to hear from you!

Do you have a suggestion for a book topic?

Let us know and it may be our next book!

www.YMBAgroup.com

TABLE OF CONTENTS

How To Use This Book

Thank you for choosing the Y.M.B.A. learning workbook series. I am excited to share the topics with you. As a teacher, corporate professional, M.B.A. and parent, I sought to find a quality program for my children that was both at an introductory level and interesting for their age. When I discovered nothing like this existed, Y.M.B.A. began. A business learning program for young students created and designed by an M.B.A, teacher, and parent. Y.M.B.A. presents information in clear, easy to follow style; focused on students approximately 12 to 16 years of age. I designed the lessons as a combination textbook and workbook because students retain far more when applying the newly taught ideas. The series instructs one idea at a time in a straightforward and simple to understand format. While presenting students with a concept they develop their understanding with fun, level-appropriate examples. After each lesson page is a worksheet to apply the idea from the page prior. This pattern keeps students engaged and actively learning with on-going student applications. The "The Drawing Board" worksheets reinforce the lesson as students practice reasoning, computation, or analysis. Y.M.B.A. focuses on useful business and everyday topics found across industries and in daily life.

Each learning workbook has a quiz for a student demonstration of their new understanding of the subject. As the student completes the learning workbook you will likely see an increase in both pride and confidence. Why wait for business concepts to be introduced? Students are ready to learn about practical life and business topics today. Y.M.B.A. lessons include relevant examples based on familiar student scenarios to sustain learning that is both effective and fun!

Business skills are useful in every industry; an understanding of business is essential. Students can begin achieving more with Y.M.B.A. today and build a path for the future. Your support is appreciated. Suggestions, questions, or comments are always welcome.

Thank you,

L.J. Keller

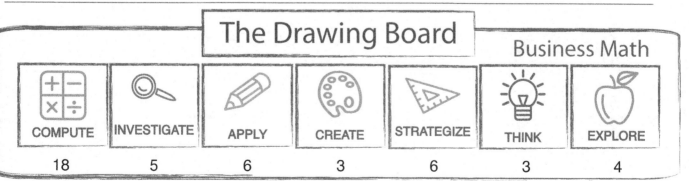

The quantity of each skill practice area is shown below each learning tile.
Worksheet pages seek to capture student interest and build learning momentum.

Fraction To Percent

Divide $5 \overline{)1.00} = .20$
change .20 to a percent
by moving the decimal
right two places.
.20 = 20%

To change a percent to a fraction
place the percent over 100 and then
reduce the fraction to the lowest terms.

$18\% = \dfrac{18}{100} = \dfrac{9}{50}$

Reduce To The Lowest Terms

1. $\dfrac{1}{4}$ = 25 %

2. $\dfrac{4}{20}$ = ___ %

3. $\dfrac{10}{70}$ = ___ %

4. $\dfrac{3}{6}$ = ___ %

5. $\dfrac{1}{8}$ = ___ %

6. $\dfrac{10}{50}$ = ___ %

7. $\dfrac{14}{16}$ = ___ %

8. $\dfrac{5}{30}$ = ___ %

9. $\dfrac{5}{8}$ = ___ %

10. $\dfrac{9}{36}$ = ___ %

11. $25\% = \dfrac{25}{100} = \dfrac{1}{4}$

12. 40% = ___

13. 82% = ___

14. 75% = ___

15. 30% = ___

16. 50% = ___

17. 64% = ___

18. 17% = ___

19. 12.5% = ___

COMPUTE

www.YMBAgroup.com

The Drawing Board

Consider the equation:

$$10 \div 2 - 1 + 3 \times 4 =$$

Consider this wrong answer:

What if you work from left to right?
10 divided by 2 equals 5.
5 minus 1 equals 4.
4 plus 3 equals 7.
7 times 4 equals 28.

**But 28 is not the correct answer.
So, where do you begin?**

Luckily in math there is a guideline called PEMDAS that directs a person solving a problem in the steps to compute the correct answer. The first step is to complete the part of the equation in the parenthesis (if any). Next, calculate the exponents (if any), then moving from left to right, multiply or divide (if any). The last step is addition or subtraction completed by moving from left to right in the equation (if any).

Therefore, in the example above the correct answer is:

Step 1: There are no parenthesis.

Step 2: There are no exponents.

Step 3: 10 divided by 2 equals 5, 3 multiplied by 4 is 12
   The new formula is now: 5 - 1 + 12 =

Step 4: Add left to right, 4+12 = 16   **16** ✓

P	E	[M	D]	[A	S]
Parenthesis	Exponents	Multiplication	Division	Addition	Subtraction

Solve The Equations Below Using PEMDAS

1. $21 \div 7 + 3 - 4 \times 2 =$

2. $(5 \times 4) + 6 - 3 =$

3. $5^2 + 3 =$

4. $80 - (5 + 4) \times 8 =$

5. $25 \div 5 \times 4 + 2 =$

6. $(3 + 2) \times 4 \times 3 =$

7. $(9 \times 4) + 6 - 3 =$

8. $3^3 + 3^3 =$

STRATEGIZE

 www.YMBAgroup.com

The Drawing Board

Savings Accounts

Saving money, even a little at a time, can add up to an investment. Complete the chart below to show the monthly, quarterly, and annual balance for each investment.

Month = 1 month	Quarter = 3 months	4 Quarters = 1 year
Monthly Savings Amount	Savings Balance After 1 Quarter	Savings Balance After 1 Year
1. $25	$25 x 3 = $75 _____	$75 x 4 = $300 _____
2. _____	$80	_____
3. _____	_____	$200
4. $50	_____	_____
5. _____	$110	_____
6. _____	_____	$160
7. $45	_____	_____
8. _____	$60	_____
9. _____	_____	$1200
10. $10	_____	_____
11. _____	$500	_____
12. _____	_____	$220

13. What would be the total saved in one month by saving $1.00 a day, plus $5 a week and $10 additional on days 10 and 20 in a month with 30 days.

$ _____

COMPUTE

www.YMBAgroup.com

The Drawing Board

A bank checking account is used to write a check so the account holder can send money in the mail or when a large dollar amount is paid. Consider the transactions that follow and complete the running balance. A running balance is the total that would be in the checking account at the bank after the check amount is deducted (taken out) from the account.

Opening deposit entered, followed by 7 check purchases.

	Running Balance
1. Shaye's Shady Day Beach Umbrella Stand opens a checking account with a $1,000 deposit into the account.	$1,000
2. That same day a check is written for $125 to a supplier.	
3. The company also writes a check to the electric company, $54.34	
4. Next week a check is written to buy an outdoor sign, $118.40.	
5. Due to high demand 5 umbrellas are bought for inventory, $50 each.	
6. A check is mailed to the telephone company, $68.72.	
7. A check is written at the office supply store, $82.90.	
8. On the last day of the month a booth rent check is mailed, $212.	

9. Based on the check written to the electric company, plus the check written to the telephone company, (both for one month of service), what can the company estimate will be the annual (12 month) electric and telephone expenses total cost?

THINK

The Drawing Board

A Check Register

Enter the transactions below to the check register. The check numbers will go in order starting at 1112. The first two are completed as examples.

1. Monthly tennis payment due, $45.50.
2. Receive your $5 weekly allowance.
3. Visit a doctor and pay $15.
4. Mail a car payment, $185.
5. Purchase a book for $8.75.
6. You give a friend $60.00.
7. Sell 5 comic books at $1 each.
8. Washed neighbors car, $5.
9. Buy $21.38 of groceries, pay by check.
10. Gather your coins and deposit $7.45.
11. Receive a $30 birthday check.
12. Monthly interest from bank, .82 cents.

	NUMBER	DATE	TRANSACTION	WITHDRAWAL	✓	DEPOSIT	624 00
1.	1112	8/2	American Tennis Club	45 50			578 50
2.	--	8/6	Weekly Chore Allowance			5 00	583 50
3.		8/9					
4.		8/12					
5.		8/15					
6.		8/16					
7.		8/18					
8.		8/21					
9.		8/23					
10.		8/27					
11.		8/29					
12.		8/31					

APPLY

www.YMBAgroup.com

The Drawing Board

Write A Check

The previous page asked you to record 12 transactions from the checking account. Below practice writing checks for the items numbered 1, 3 and 4 shown on the prior page. Use the completed clues to help complete each check.

Check Stub

Checks To Write

Date	
October 7, 2019	

Company
American Tennis

Amount
$45.50

Memo
October
Tennis

YMBA Student
4 Success Blvd.
Beautiful, State 98765 USA

58-6499/5758   1112

Date _____

Pay To The Order Of: __American Tennis_____ $ | $45.50 |

_____ Dollars

America United Bank

memo: _____   Signature: _____

I: 575864996■:   II■ 758969998II   1112

Date
October 14, 2019

Company
Dr. Fixmeup

Amount
$15.00

Memo
Doctor
Visit

YMBA Student
4 Success Blvd.
Beautiful, State 98765 USA

58-6499/5758
1113

Date 10/14/2019 _____

Pay To The Order Of: _____ $ [   ]

Fifteen dollars and zero cents _____ Dollars

America United Bank

memo: _____   Signature: _____

I: 575864996■:   II■ 758969998II   1113

Date
October 18, 2019

Company
MYE Bank

Amount
$185.00

Memo
October Car
Payment

YMBA Student
4 Success Blvd.
Beautiful, State 98765 USA

58-6499/5758   1114

Date _____

Pay To The Order Of: _____ $ [   ]

_____ Dollars

America United Bank

memo: __October Car Payment__   Signature: _Your M_

I: 575864996■:   II■ 758969998II   1114

CREATE

Grocery Shopping

7 ¢/ounce
Grapes

Bananas
$3
1/2 dozen

Tomatoes
$1.00/Lb.

1 Dozen
Apples
$5.00

Plums
2 / $1

Mangos
80 ¢ each

40 cents
per pound
Lemons

2 Totes
Strawberries
$8

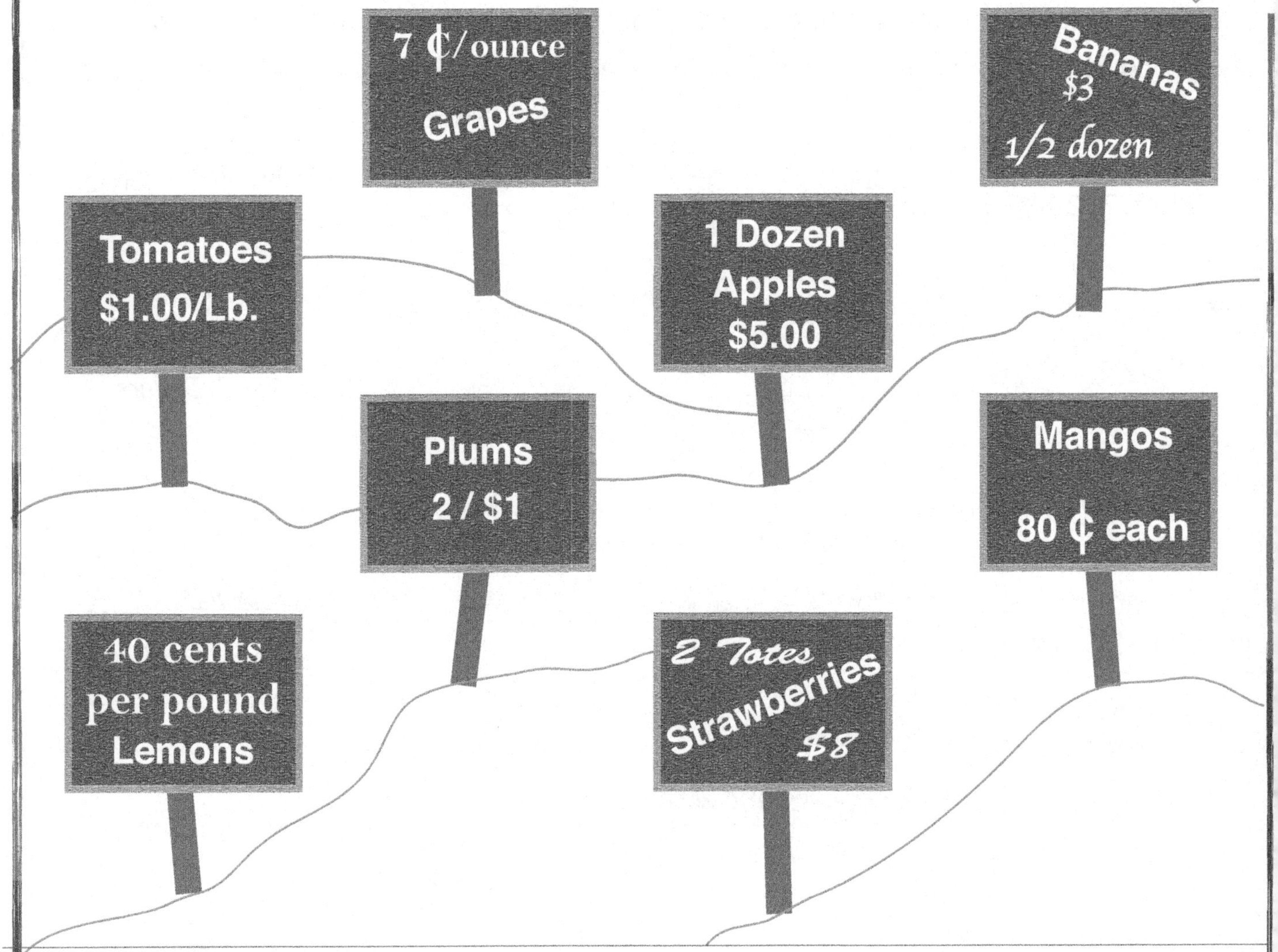

Customer 1	Customer 2	Customer 3
4 mangos	1 tote strawberries	2 pounds lemons
6 bananas	10 oz. grapes	2 dozen bananas
3 pounds lemons	2 plums	4 plums
1 pound tomatoes	1 mango	2 totes strawberries
8 ounces grapes	3 lbs. tomatoes	12 apples
$	$	$

4. Which customer spent the most money today? _____

STRATEGIZE

The Drawing Board

You just finished shopping and your items total cost is $28.45. The cashier will begin to deduct your coupons below from your total. What is the total amount to pay after the coupons are deducted from your total?

You purchased:

2 bags of Saltz Pretzels   1 bottle Bubble-Clean Hand Soap
1 can of Rodeo Baked Beans   1 box of Healthy Eats Cereal
1 bottle of Yumamin Water   2 boxes Amore Frozen Pizza
1 16 oz. package USA Cheese   1 4 pack of Ever-Charge Batteries

$1 off per customer
Pretzels
Save $1
Per Customer

Save 80 cents
Hand Soap

$.65
Off
Cereal

25 cents off Cheese
Yum

Baked Beans save
$0.50
per can

water
40 cents off one

$1 off 2
Pizza
Save $1

Batteries
$1.50
per package

1. Grocery Amount Due After Coupons: $ _____

COMPUTE

www.YMBAgroup.com

A Super Sale

Today you are shopping at your favorite clothing store.
You decide to buy 2 pairs of jeans, 1 sweater, 5 shirts and 8 pairs of socks.
The state you are in will add 6.5% sales tax on all clothing purchases.
What is the total amount spent?

Shirts $12 Each

Sweaters $20 Each

Jeans $40 Buy 1 Get 1 Free!

Socks and Gloves $2

1. Shirts Total: _____

2. Sweaters Total: _____

3. Jeans Total: _____

4. Socks Total: _____

5. Gloves Total: _____

6. Sub-Total: _____

7. Sales Tax: _____

8. Sub-Total plus Sales Tax: _____

9. Total Amount Spent: _____

10. Paid $160, Change Received: _____

APPLY

www.YMBAgroup.com

The Drawing Board

Sales Tax

Often when you make a purchase a store will have to collect an additional percentage of the price to give to the state. Compute the sales tax amount below then compute the change received after payment.

1. Purchase Price: $58.22

   $58.22
   x   .05
   $2.91

   State Sales Tax Rate: 5%

   Total Due: $ $58.22+$2.91=$61.13

   Total Cash Given: $65.00

   Change Received: $ $65.00-$61.13=$3.87

2. Purchase Price: $114.82

   State Sales Tax Rate: 7%

   Total Due: $ _____

   Total Cash Given: $150.00

   Change Received: $ _____

3. Purchase Price: $85.00

   State Sales Tax Rate: 6%

   Total Due: $ _____

   Total Cash Given: $100.00

   Change Received: $ _____

4. Purchase Price: $253.62

   State Sales Tax Rate: 8.5%

   Total Due: $ _____

   Total Cash Given: $300.00

   Change Received: $ _____

5. Purchase Price: $31.54

   State Sales Tax Rate: 6.2%

   Total Due: $ _____

   Total Cash Given: $40.00

   Change Received: $ _____

6. Purchase Price: $65.25

   State Sales Tax Rate: 8%

   Total Due: $ _____

   Total Cash Given: $80.00

   Change Received: $ _____

7. Purchase Price: $124.00

   State Sales Tax Rate: 6.5%

   Total Due: $ _____

   Total Cash Given: $140.00

   Change Received: $ _____

COMPUTE

www.YMBAgroup.com

The Drawing Board

Recipe Conversions

Delicious! You decide to go shopping to prepare your favorite cookie recipe for 16 people today. You will also shop to bake cookies for 40 people next week. Compute the conversions to find the quantity needed of each ingredient.

GRANNY G'S SUGAR COOKIES*
1 POUND BUTTER
2 CUP SUGAR
3 EGGS
1/2 TABLESPOON LEMON JUICE
2 TEASPOONS CINNAMON
6 CUPS FLOWER
1 TEASPOON BAKING POWDER
SERVES 8 PEOPLE A TOTAL OF 16 COOKIES

What would be your quantities of each ingredient if serving 16 people?
1. ___2___ pounds butter
2. _____ cups sugar
3. _____ eggs
4. _____ tablespoons lemon juice
5. _____ teaspoons cinnamon
6. _____ cups flour
7. _____ teaspoons baking powder
8. Serves _____ people a total of _____ cookies.

What would be your quantities of each ingredient if serving for 40 people?

9. _____ pounds butter
10. _____ cups sugar
11. _____ eggs
12. _____ tablespoons lemon juice
13. _____ teaspoons cinnamon
14. _____ cups flower
15. _____ teaspoons baking powder
16. Serves _____ people a total of _____ cookies.

*This is not a real recipe. Shown only for conversion practice.

EXPLORE

www.YMBAgroup.com

The Drawing Board

Mortgage Payments

When buying a home some people choose to request a mortgage from the bank. A mortgage is a loan to buy a home. The bank will review the application details including current employment information and the applicant's credit report before deciding if they will lend the money. If the bank chooses to loan the applicant the money to purchase a home the bank will also charge interest. Interest is most commonly a percentage of the total loan amount.

Simple formula to compute 1 year of interest:
Interest = Loan Amount x (Interest Rate x Time)

Consider the scenarios below. Then complete the total amount owed by the borrower over the term of the mortgage using each of the different interest rates.

1. Mortgage Loan Amount: $184,000
   Annual Interest Rate: 3%
   30 Year Mortgage
   Total Interest In 1 Year: $__5,520__

   $184,000x(.03x1)=$5,520

2. Mortgage Loan Amount: $345,000
   Annual Interest Rate: 4.8%
   30 Year Mortgage
   Total Interest In 1 Year: $_____

3. Mortgage Loan Amount: $275,000
   Annual Interest Rate: 5%
   15 Year Mortgage
   Total Interest In 1 Year : $_____

4. Mortgage Loan Amount: $184,000
   Annual Interest Rate: 7%
   30 Year Mortgage
   Total Interest In 1 Year: $_____

5. Mortgage Loan Amount: $345,000
   Annual Interest Rate: 6.2%
   30 Year Mortgage
   Total Interest In 1 Year: $_____

6. Mortgage Loan Amount: $275,000
   Annual Interest Rate: 8%
   15 Year Mortgage
   Total Interest In 1 Year : $_____

7. How does a different interest rate effect the amount of interest paid?

... Did you know the loan amount is also known as 'principal'?

COMPUTE

www.YMBAgroup.com

The Drawing Board

Complete the chart below to show the computed rent amounts.

1 year = 365 days (12 months)   4 weeks = 1 month   1 week = 7 days

	(A) Daily	(B) Weekly	Monthly	(C) Annual
1.	$600 ÷ 7 = $85.71	$2,400 ÷ 4 = $600	$2,400	$2,400 x 12 = $28,800
2.			$ 900	
3.			$1,200	
4.			$1,750	
5.			$1,928	
6.			$1,435	
7.			$1,600	
8.			$ 650	
9.			$ 810	
10.			$1,410	
11.			$2,000	
12.			$1,150	

13. What are other monthly expenses related to an apartment?

COMPUTE

                                       www.YMBAgroup.com

The Drawing Board

Consider This Scenario:

Rental Living Expenses

Move-In Security Deposit, $1,000 one-time payment.
Rental Payment $900 per month, $10,800 per year.
Rental Insurance $20 per month, $240 per year.
Water Bill $38 per month, $456 per year.
Electric Bill $89 per month, $1,068 per year.
Telephone Bill $65 per month, $780 per year.

Decision To Move:

30 days notice given to owner.
$1,000 security deposit returned since no damage.
Cash In Hand after 1 year = $1,000

1. What is the total monthly cost to rent the apartment?

Homeowner Living Expenses

Closing Costs To Obtain A Mortgage, $4,000 one-time payment.
Mortgage Payment $1,000 per month, $12,000 per year.
Homeowners Insurance $50 per month, $1,100 per year.
Water Bill $38 per month, $456 per year.
Electric Bill $89 per month, $1,068 per year.
Telephone Bill $65 per month, $780 per year.

Decision To Move:

Home Listed With A Real Estate Agent, 90 days until sold.
$3,000 paid to real estate agent at time of home sale.
Real Estate Market Has Usual Growth, Home Value Increases $23,000
Cash In Hand after 1 year = $4,000

2. What is the total monthly cost to own the home?

INVESTIGATE

Invoices

Review the invoice below given to a customer at a retail store. Remember to examine the invoice for hidden clues in the stock numbers.

Perfect Party Rentals, Incorporated

INVOICE NUMBER: 5369
DUE DATE: UPON PICK-UP

March 17, 2019

Quantity	Stock Number	Item Details	Sales Rep Initials	Total Price
18	CHWHITE	Standard party chair	MBA	$107.10
4	TABLE4X6	Standard party table	MBA	$100.00
25	BALLG	Large balloons	MBA	$ 28.75

Sub-Total	$235.85
State Sales Tax 7%	$16.51
Total	$252.36

Perfect Party Rentals- PPE Delivers
24/7
LOCATION
ORLANDO, FL
ANY TIME
EVERY DAY

1. What is the invoice number and date of the transaction?

2. What type of company issued the invoice?

3. Where is the company located?

4. How many chairs did the customer request?

5. What is the cost per chair?

6. How many balloons did the customer request?

7. What is the cost per balloon?

8. Sales tax shown is 7%. What would be the sales tax cost at 4%?

9. The customer pays with $300, what is the change given back?

10. Bonus: What color are the chairs?

STRATEGIZE

Credit Cards

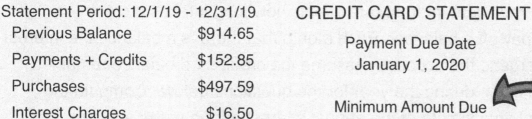

CREDIT CARD STATEMENT

Statement Period: 12/1/19 - 12/31/19	
Previous Balance	$914.65
Payments + Credits	$152.85
Purchases	$497.59
Interest Charges	$16.50
Current Balance	$514.09

Payment Due Date
January 1, 2020

Minimum Amount Due
$52

Reminder: Always Pay More Than The Minimum Amount Due To Pay Less Interest.

12/04/19	Get Gas Here Company	Atlanta, GA	$42.68
12/05/19	Late Fee		$25.00
12/09/19	Payment Received		- $150.00
12/11/19	Fire Department Holiday Raffle		$20.00
12/15/19	Wild Wrapping Paper	Savannah, GA	$12.43
12/16/19	Doggie Bath Stop	Central GA	$35.25
12/16/19	Fresh Foods Grocery	Foodtown, GA	$85.76
12/18/19	Get Gas Here Company	Atlanta, GA	$31.14
12/23/19	Little Box Gift Shop	Main Village	$216.86
12/26/19	Return: Wild Wrapping		- $2.85
12/31/19	www.PartyHatsOFun.com		$28.47
12/31/19	Interest Charges		$16.50

1. Review the transactions. In what state does it appear the cardholder lives?

2. How much is the late fee charged for not having a payment made on time?

3. What is the amount of the payment that was made in December?

4. Which one item on the transaction list was purchased on the computer?

5. What two locations were visited on the same day?

6. How much interest was added as a charge on the account?

7. Explain the transaction that happened on December 26, 2017.

8. The credit card owner most likely has what type of pet?

9. What is the total amount of gas charges in December?

INVESTIGATE

The Drawing Board

Credit Card Interest

The amount of a credit card interest rate effects how long it takes a credit card customer to pay off a balance. Each month that there is a balance the interest rate will add a charge to the account. Assume the credit card balance remained approximately the same during the year for the questions below. Compute the scenarios to reveal an estimate of the annual interest added to the account.

1.
Credit Card Balance: $2,654
Annual Interest Rate: 12%
Estimated Interest Added To The Account: $ ___ $2,654 x .12 = $318.48

2.
Credit Card Balance: $5,328
Annual Interest Rate: 8%
Estimated Interest Added To The Account: $ _____

3.
Credit Card Balance: $12,125
Annual Interest Rate: 2.5%
Estimated Interest Added To The Account: $ _____

4.
Credit Card Balance: $8,753
Annual Interest Rate: 7%
Estimated Interest Added To The Account: $ _____

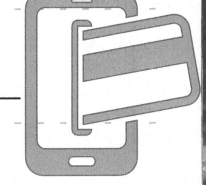

5.
Credit Card Balance: $512
Annual Interest Rate: 5%
Estimated Interest Added To The Account: $ _____

6. Which account had the highest card balance? _____

7. Which account balance had the most interest added? _____

8. Why did the account with the highest balance not have the most interest added? _____
_____

COMPUTE

www.YMBAgroup.com

The Drawing Board

An exponent shows how many times to multiply the same number.

For example, 2^5 equals 2x2x2x2x2.

Solve the exponents below.

$2 \times 2 \times 2 \times 2 \times 2 =$

$4 \times 2 \times 2 \times 2 =$

$8 \times 2 \times 2 =$

$16 \times 2 =$

$2^5 = 32$

1. $2^5 =$ 2x2x2x2x2=32

2. $3^4 =$

3. $5^2 =$ 10. $0^7 =$

4. $8^3 =$ 11. $3^5 =$

5. $3^7 =$ 12. $6^2 =$

6. $4^5 =$ 13. $2^6 =$

7. $1^5 =$ 14. $11^2 =$

8. $9^1 =$ 15. $7^3 =$

9. $7^3 =$ 16. $6^3 =$

COMPUTE

25

Square Roots

Finding the square root of a number is completed by finding the number that when multiplied by itself equals the number under the radical sign.

Radical Sign ➔ $\sqrt{16} = 4^2 = 4 \times 4$

To solve a square root, ask yourself ...

what number when multiplied by itself equals the number under the radical.

Solve A Square Root

1. $\sqrt{64}$ = ___ X ___

2. $\sqrt{121}$ = ___ X ___

3. $\sqrt{9}$ = ___ X ___

4. $\sqrt{49}$ = ___ X ___

5. $\sqrt{144}$ = ___ X ___

6. $\sqrt{36}$ = ___ X ___

7. $\sqrt{81}$ = ___ X ___

8. $\sqrt{4}$ = ___ X ___

9. $\sqrt{400}$ = ___ X ___

10. $\sqrt{225}$ = ___ X ___

Make A Square Root

11. 5 X 5 = $\sqrt{}$

12. 1 X 1 = $\sqrt{}$

13. 10 X 10 = $\sqrt{}$

14. 7 X 7 = $\sqrt{}$

15. 25 X 25 = $\sqrt{}$

16. 8 X 8 = $\sqrt{}$

17. 12 X 12 = $\sqrt{}$

18. 2.5 X 2.5 = $\sqrt{}$

COMPUTE

The Drawing Board

Manufacturing

Stinkly La Peu Peu has been experiencing an increase in consumer demand. An increasing number of people are buying the perfume. The company is considering making changes to the size of each bottle. Help the company analyze data as part of considering the bottle size change.

Bravo   *Bien! Bien!*

S.L.P.P.
8 oz.

1. Current production manufactures 96 ounces per hour of perfume. How many Stinkly Le Peu Peu perfume bottles are filled per hour? _____

2. If the manufacturing plant operates 10 hours a day how many bottles are manufactured each day? _____

3. The company may change the package so that each perfume bottle holds 5 ounces. With the change to 5 ounces how many bottles would the company manufacture per hour? _____   In ten hours? _____

4. The perfume is currently sold at $3 per ounce. What is the perfume price at the current size of 8 ounces? $_____

5. What would be the perfume price at the size of 5 ounces? $_____

6. The current cost to manufacture the perfume per ounce is $1.40. What is the perfume manufacturing cost for an 8 ounce bottle? $_____

7. What is the perfume manufacturing cost for the 5 ounce bottle? $_____

8. The profit is found by subtracting the manufacturing costs from the retail price. What is the profit for each 8 ounce perfume bottle? $_____
   What is the profit for each 5 ounce perfume bottle? $_____

9. Which perfume bottle size yields (earns) the highest profit?

Spectacular   *Oh La La*   _____

THINK

www.YMBAgroup.com

The Drawing Board

Efficiency and Time

Time is money. A common phrase, but what does it mean?
In business manufacturing the phrase refers to the need to have a production line that produces products efficiently. By having an efficient production line each item is produced in the fastest amount of time possible. In business each minute, hour or full day of work results in expenses that are to be paid. Operating an efficient business requires the production of items of the highest quality in the least amount of time possible.

ASSEMBLY LINE

Squizzle-Star Production

| 30 seconds | 30 seconds | 1 minute | 35 seconds | 25 seconds |

1. What is the total production time for one squizzle-star? _____

2. What step in the production line takes the greatest amount of time? _____

3. How many employees are on the squizzle-star production line? _____

4. How many squizzle-star toys are produced in one hour? _____

5. How many squizzle-star toys are produced in 8 hours? _____

INVESTIGATE

www.YMBAgroup.com

The Drawing Board

Profit Calculations

A retail business commonly purchases goods to sell to customers from a type of business known as a wholesaler. A wholesaler will purchase from a company that manufacturers the goods. The retail business will present the goods for sale and add an amount on to the price they paid to the wholesaler. This new price is the retail sales price. **The difference between the retail sales price and the price paid the wholesaler is the company profit.** (subtract)

Compute the blanks in the chart below then calculate the total profit per item.

1. Wholesale Cost Per Item: $ __2.45__
2. Total Quantity Purchased: __100__
3. Total Wholesale Cost: $ _____

4. Retail Price Per Item: $ _____
5. Total Quantity Sold: __83__
6. Total Retail Sales: $ __415__

6. Total Retail Sales: $ __415__
7. Total Wholesale Cost: $ _____
8. Total Item Profit: $ _____

9. Wholesale Cost Per Item: $ _____
10. Total Quantity Purchased: __70__
11. Total Wholesale Cost: $ __1,050__

12. Retail Price Per Item: $ __26__
13. Total Quantity Sold: __63__
14. Total Retail Sales: $ _____

15. Total Retail Sales: $ _____
16. Total Wholesale Cost: $ __1,050__
17. Total Item Profit: $ _____

18. Wholesale Cost Per Item: $ _____
19. Total Quantity Purchased: __2,400__
20. Total Wholesale Cost: $ __8,160__

21. Retail Price Per Item: $ __6.20__
22. Total Quantity Sold: __1,754__
23. Total Retail Sales: $ _____

24. Total Retail Sales: $ _____
25. Total Wholesale Cost: $ __8,160__
26. Total Item Profit: $ _____

COMPUTE

The Drawing Board

Depreciation

When an item is purchased the starting value is close to the amount paid. However, over time the value of an item decreases. For this reason, accounting uses depreciation to have an accurate value for the assets of a business. Depreciation takes into consideration that the actual value of an asset decreases over time. The decrease in value may be caused by normal use, or the item getting older, or the general usefulness of the item being reduced as compared to new versions. For example, a computer purchased for $3,000 in 2005 is no longer worth $3,000 today, but why? ... The computer was used, so it is no longer new, plus updated versions that are more useful and efficient are available.

Straight Line Depreciation

Purchase Price		10 Years Useful Life		Value in 10 Years

$3,000 $2,700 $2,400 $2,100 $1,800 $1,500 $1,200 $900 $600 $300 $0

Practice straight-line depreciation in the examples below.

1. $28,700   7 years   Annual depreciation to bring asset to $0: $__$4,100__

   $28,700 ÷ 7 = $4,100 per year

2. $12,000   6 years   Annual depreciation to bring asset to $0: $_____

3. $3,500   5 years   Annual depreciation to bring asset to $0: $_____

4. $25,000   10 years   Annual depreciation to bring asset to $0: $_____

5. $18,900   6 years   Annual depreciation to bring asset to $0: $_____

6. $140,335   5 years   Annual depreciation to bring asset to $0:

   $_____

COMPUTE

The Drawing Board

Car ownership is a common goal in the United States. However, the cost to own a car is often greater than expected. At times a car loan may be chosen to help pay the car. Consider the questions below that compare two different scenarios regarding different car loans over a length of time.

Year To Months Calculations

1. How many months are in 1 year? _____

2. How many months are in 3 years? _____

3. How many months are in 5 years? _____

4. How many months are in 7 years? _____

Which car loan costs the buyer more?

5. 5 year loan, monthly payment $217   7 year loan, monthly payment $185

    $60 \times \$217 = \$13,020$ $84 \times \$185 = \$15,540$

6. 3 year loan, monthly payment $420   5 year loan, monthly payment $374

7. 7 year loan, monthly payment $165   3 year loan, monthly payment $298

8. 5 year loan, monthly payment $308   7 year loan, monthly payment $250

9. 3 year loan, monthly payment $512   7 year loan, monthly payment $382

10. Which payment plan would you prefer?

11. Why would a buyer choose a 7 year plan rather than a 3 year plan?

STRATEGIZE

31

The Drawing Board

Car Repairs

A chart has been created below to track the various expenses paid by the owner of a car. Review the chart below then consider questions one to four that follow below.

Jan	Feb	Mar	
Car payment, $172 Gas, $65 Car Insurance, $209	Car payment, $172 Gas, $29 Car wash, $12	Car payment, $172 Gas, $29	1st Quarter
Apr Car payment, $172 Gas, $58 Car Insurance, $209 Oil change, $36	**May** Car payment, $172 Gas, $47 New tires, $385	**Jun** Car payment, $172 Gas, $60 Car wash, $12	2nd Quarter
Jul Car payment, $172 Gas, $42 Car Insurance, $209	**Aug** Car payment, $172 Gas, $67 Oil change, $36	**Sep** Car payment, $172 Gas, $54 New wiper blades, $24	3rd Quarter
Oct Car payment, $172 Gas, $32 Car Insurance, $209	**Nov** Car payment, $172 Gas, $48 Car wash, $12	**Dec** Car payment, $172 Gas, $72 Oil change, $36	4th Quarter

1. What is the average cost, per month, to own a car in the 3rd quarter?

2. What is the total cost in the 3rd quarter to own a car in the example above?

3. What is the total cost to own a car for the year in the example above?

4. If the car owner above earns $14 per hour, how many hours are needed to work to pay for the 3rd quarter car expenses?

INVESTIGATE

www.YMBAgroup.com

The Drawing Board

CARPET INSTALL MEASUREMENTS

All Area Carpets, Inc. has a full schedule today. Ten customers have requested measurements and price quotes for their homes. Assist the company by completing the following questions.

Cost

4. Room 10x18, $5.00 sq. ft. $ _____

5. Room 11x20, $4.25 sq. ft. $ _____

6. Room 19x24, $3.10 sq. ft. $ _____

7. Room 14x16, $6.35 sq. ft. $ _____

8. Room 14x22, $5.42 sq. ft. $ _____

9. Room 16x31, $4.00 sq. ft. $ _____

10. Room 8x15, $11.25 sq. ft. $ _____

Rectangle

7

3

$5 SQ. FT.

1.
Square Feet: __3 x 7 = 21__

Cost To Install: $ __21 x 5 = $105__

Rectangle

$3 SQ. FT.

2. Square Feet: _____

Cost To Install: $ _____

4

9

Rectangle

$4 SQ. FT.

3

7

5

2

3. Square Feet: _____

Cost To Install: $ _____

APPLY

www.YMBAgroup.com

The Drawing Board

BASE FLOOR MOLDING INSTALL MEASUREMENTS

Peri's Perfect Trim Molding Company schedules all base trim molding measurements on the same today of the week. Base trim molding is installed around the room where the wall meets the floor. Assist Peri by completing quote requests below.

1. Room 11x14, $2 per linear foot $ ___100___

    $22 + 28 = 50 \quad 50 \times \$2 = \$100$

2. Room 15x10, $3 per linear foot $ _____

3. Room 12x24, $4 per linear foot $ _____

4. Room 20x11, $5 per linear foot $ _____

5. Room 14x18, $6 per linear foot $ _____

6. Room 5x7, $7 per linear foot $ _____

7. Room 6x10, $8 per linear foot $ _____

Square

$3 LINEAR FOOT

3

8. Linear Feet: _____

   Cost To Install: $ _____

Square

$2 LINEAR FOOT

4

9.
Linear Feet: _____

Cost To Install: $ _____

Rectangle

10. Linear Feet: _____

   Cost To Install: $ _____

$4 LINEAR FOOT

4

5

Rectangle

11. Linear Feet: _____

   Cost To Install: $ _____

3

7

5

2

$2 LINEAR FOOT

APPLY

Travel Time

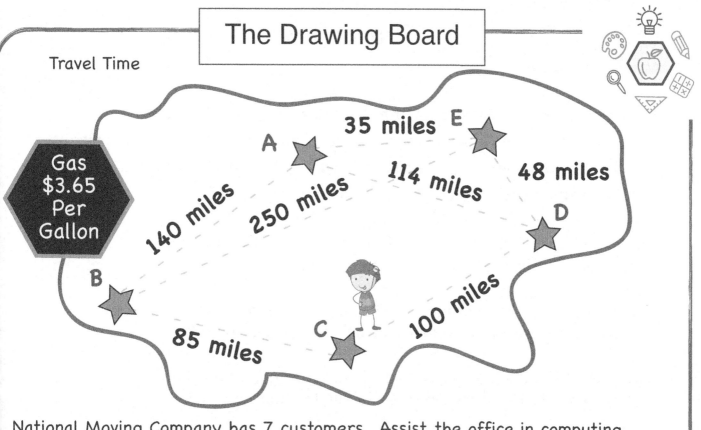

Gas $3.65 Per Gallon

35 miles   E

A

114 miles   48 miles

D

140 miles   250 miles

B

100 miles

85 miles   C

National Moving Company has 7 customers. Assist the office in computing how much gas each truck driver will need for the 5 routes below.

Note: Each truck travels 25 miles per each gallon of gas.

1. Route 1: Departing A and delivering at D.

Total Miles: _____   Total Gas Gallons Needed: _____

Total Gas Cost Route 1: $ _____

2. Route 2: Departing B, passing by C and delivering at D.

Total Miles: _____   Total Gas Gallons Needed: _____

Total Gas Cost Route 2: $ _____

3. Route 3: Departing D, picking up at A and delivering at E.

Total Miles: _____   Total Gas Gallons Needed: _____

Total Gas Cost Route 3: $ _____

4. Route 4: Departing E, making a partial delivery at A and delivering at B.

Total Miles: _____   Total Gas Gallons Needed: _____

Total Gas Cost Route 4: $ _____

COMPUTE

The Drawing Board

The large percentage of cars on the road use gasoline for energy. Gasoline when sold by a gas station is measured in gallons. The gas price is shown to the customer as per gallon. A car that is said to be 'fuel-efficient' drives further on each gallon of gas than a less fuel efficient car.

1. Car A has 15 gallons of gas in a full tank. If gas costs $3.64 per gallon, what is the total gas cost to fill the tank?   15 x $3.64 = $54.60

2. Car A can travel an average of 30 miles per gallon. On a full tank of gas approximately how many miles can the car travel?

3. Car B has 18 gallons of gas in a full tank. If gas costs $3.52 per gallon, what is the total gas cost to fill the tank?

4. Car B is able to travel an average of 25 miles per gallon. On a full tank of gas approximately how many miles can the car travel?

5. Car C left the gas station with a full tank of gas. The car traveled three hundred miles and ran out of gas. If the car gets an average of 25 miles per gallon, how many gallons does the car hold when its gas tank is full?

6. Car D arrived at the gas station and filled the car with gas. If the car added 12 gallons of gas and the cost per gallon is $3.20, what was the total amount spent on gas?

7. Which is a more efficient car: Car 1 travels 300 miles on a full tank of gas. Car 1 holds 16 gallons of gas in a full tank. Car 2 travels 240 miles on a full tank of gas and holds 12 gallons of gas on a full tank.

8. Which car can travel further? Car E has a full tank of gas and can travel 21 miles per gallon on a full 14-gallon tank. Car F has a full tank of gas and can travel 29 miles per gallon on a full 11-gallon tank.

9. Which car can travel further? Car G has a full tank of gas and can travel 28 miles per gallon a full 15-gallon tank. Car H has a full tank of gas and can travel 36 miles on a full 12-gallon tank.

STRATEGIZE

The Drawing Board

Salary

Employment earnings that are paid as a salary are given to the employee without a strict tracking of the hours work started and stopped. When paid with a salary the employee is expected to complete the tasks of their job and some days may have more work hours than others required to complete work. Compute the hourly wages below.

	Annual Salary	Monthly Earnings	Weekly Earnings	Weekly Net Earnings After 21% Taxes
1.	$42,000 divide by 12	$3,500 divide by 4	$875	$875 - ($875x.21) = $691.25
2.	$65,000			
3.	$25,000			
4.	$35,000			
5.	$82,000			
6.	$142,000			
7.	$18,000			
8.	$20,250			
9.	$63,930			
10.	$36,240			

COMPUTE

The Drawing Board

Employment earnings may be paid hourly or as a salary.

Compute the hourly wages and deduct the taxes to find the net earnings.

	Hourly Rate	Weekly Hours Worked	Deduct 21% Taxes	Net Earnings
1.	$14.00	40	$117.60	$560 - $117.60 = $442.40
2.	$11.00	36		
3.	$9.50	34		
4.	$10.15	32		
5.	$9.00	18		
6.	$17.45	20		
7.	$16.20	31		
8.	$21.15	37		
9.	$26.42	40		
10.	$17.65	19		
11.	$12.84	20		
12.	$11.95	35		

COMPUTE

Payroll Taxes

Imagine you are the owner of a business. You have four employees. Listed below are the gross (before taxes) earnings for each person. Also shown are simple payroll tax deduction percentages. Compute the tax amounts to be deducted from the weekly payroll for each employee. Taxes deducted will be paid by the company to the government for the employee.

Weekly Gross (before taxes) Payroll:

Employee A	Employee B	Employee C	Employee D
$400	$440	$825	$620

Federal Government Tax	11%
State Tax Rate	7%
City Tax Rate	3%

	Federal	State	City
Employee A	$400 x .11 = $44	$400 x .07 = $28	$400 x .03 = $12
Employee B	$	$	$
Employee C	$	$	$
Employee D	$	$	$

... Did you know a company will also pay a tax approximately equal to the amount of federal tax that was paid by each employee in their paychecks? Add the total of the federal government payroll taxes paid in one week by the four employees above. Write the total below to show the payroll tax amount paid to the government as a tax on the business.

$ _____

EXPLORE

www.YMBAgroup.com

The Drawing Board

Stock Price Line Graph

Cabana Beach Resorts recently offered investors a chance to buy stock in the company for the first time. The first time stock is for sale is known as an IPO (initial public offering). The average stock price of the IPO (January) was $25.80. Additional average monthly stock prices are shown below.

FEB	$28.12	MAY	$27.00	AUG	$38.10	NOV	$38.10
MAR	$27.50	JUN	$32.48	SEP	$36.75	DEC	$39.00
APR	$24.75	JUL	$37.62	OCT	$34.00		

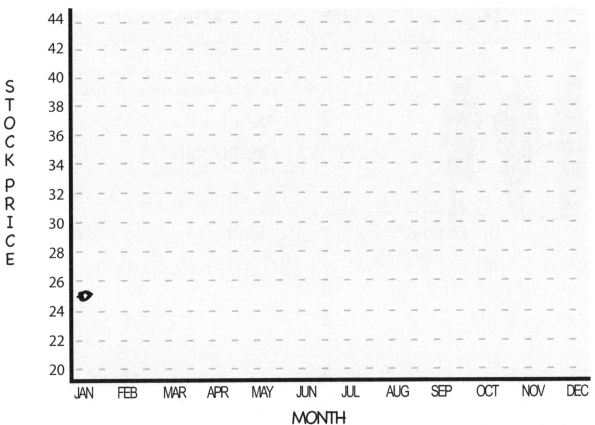

Cabana Beach Resorts 2018 Average Stock Price

Plot the average monthly stock prices on the chart above. January has been completed.
Next, connect the plot points to create a line graph to more clearly view the trend of the stock price.

1. What conclusion may be made regarding the company average stock price in 2018?

_____

_____

_____

_____

CREATE

P/E Ratio

A P/E ratio compares the price of a share of company stock to the dollar amount each single share of stock earned in dividends. A dividend is the money given to stockholders by a company. Competitor P/E ratios may be compared to gain an understanding of how a single company is performing within the industry.

$$\text{P/E Ratio} = \frac{\text{Current Stock Price}}{\text{Earnings Per Share During Last 12 Months}}$$

A High P/E Ratio = More Likely To Pay High Dividends

Compute the P/E Ratio for each company below.

Each company is the music industry.

A Low P/E Ratio = Less Likely To Pay Low Dividends

Company:	A	B	C	D
Stock Price	$54.12	$38.84	$53.80	$33.75
Earnings Per Share	$2.46	$3.28	$2.84	$1.95

Compute the P/E Ratios using the data above and answer the questions below.

1. Which company is more likely to pay a higher dividend?

2. Which company is more like to pay a lower dividend?

3. Which company is performing better than the others shown in the industry?

4. Which company is not performing well in the industry?

STRATEGIZE

www.YMBAgroup.com

A bond is purchased from a company or government by an investor.
The bond has a par value to be paid, plus interest, on the bond maturity date.

1. Social Security Number of bond owner.

2. Who will be paying interest to the owner.

3. The value of the savings bond spelled out in words.

4. The month/year when savings bond interest starts.

5. The date the savings bond was printed.

6. The bond serial number, unique to each bond.

7. The name/address of the savings bond owner.

8. The signature or name from the bond company

9. If a bond owner is under 18 a parent/guardian name.

10. The maturity value of the savings bond.

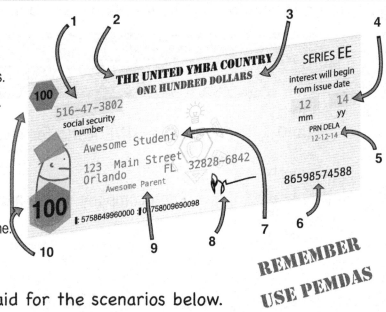

REMEMBER
USE PEMDAS

Compute the simple bond interest paid for the scenarios below.

1. $100 par value, matures in 10 years, plus 3% annual interest starting year 11.

At year 15 the value = $100 + (($100 x .03) x (15-11)) =

= $100 + $3 x 4 =

= $100 + $12   = $112

2. $200 par value, matures in 20 years, plus 5% annual interest starting year 21.

At year 25 the value = $_____ + (($ ___ x ___) x (___ - ___)) =

= $_____ + ($_____ x _____) =

= $_____ + $_____   = $_____

3. $50 par value, matures in 24 years, plus 4% annual interest starting year 25.

At year 40 the value = $_____ + (($ ___ x ___) x (___ - ___)) =

= $_____ + ($_____ x _____) =

= $_____ + $_____   = $_____

4. $500 par value, matures in 10 years, plus 2% annual interest starting year 11.

At year 28 the value = $_____ + (($ ___ x ___) x (___ - ___)) =

= $_____ + ($_____ x _____) =

= $_____ + $_____   = $_____

COMPUTE

Mutual Funds

A mutual fund is one option to invest money. A mutual fund takes one deposit from an investor and joins it with other investors deposits. The total of the investors deposits are then divided out among different investments to help reduce risk. If one part of the mutual fund does not perform well, the other parts of the fund will help balance the overall results.

LOW RISK   **MEDIUM RISK**   **HIGH RISK**

conservative   aggressive

Savings Account | Checking Account | Government Bond | Corporate Bond | IRA | Mutual Fund | Blue Chip Stock | Stock Stable Corporation | Stock New Corporation

Mutual funds can vary in the level of risk associated with each fund. Label the mutual funds below as either high risk, medium risk, or low risk.

1. 60% Bonds
    40% Mutual Funds
This is a __Medium__ risk mutual fund.

2. 80% New Company Stock
    20% Blue Chip Stock
This is a _____ risk mutual fund.

3. 40% Government Bonds
    5% Mutual Funds
    55% Checking Account
This is a _____ risk mutual fund.

4. 20% Corporate Bonds
    80% New Corporation Stock
This is a _____ risk mutual fund.

5. 75% Stable Company Stock
    25% IRA Funds
This is a _____ risk mutual fund.

THINK

www.YMBAgroup.com

The Drawing Board

Earnings Per Share

Earnings Per Share calculations are a way for investors to understand how much of a return (known as a profit) they can expect on their stock purchase in a company. To compute the earnings per share of stock the dividends being paid out first need to be subtracted from the company profit. Then the amount being paid out is divided by the number of stock shares owned by investors.

$$\text{Earnings Per Share} = \frac{(\text{Profit minus Dividends Paid})}{\text{Number of Shares Outstanding}}$$

Rapid Races

earned $875,000 in profit last year.

The company had 40,200 shares of stock outstanding.

The company paid a total of $25,380 in dividends.

1. Earnings Per Share = $

Tip-Top Tires

earned $476,000 in profit last year.

The company had 4,610 shares of stock outstanding.

The company paid a total of $15,000 in dividends.

2. Earnings Per Share = $

Lynk Luxury Autos

earned $1,068,000 in profit last year.

The company had 15,200 shares of stock outstanding.

The company paid a total of $70,000 in dividends.

3. Earnings Per Share = $

4. Imagine you are a shareholder in a corporation. Would you prefer the earnings per share number to be high or low? Why?

APPLY

The Drawing Board

After high school many students choose to continue their education at a college or university. Attending may be expensive. Often it is difficult to work a full-time job while in college since a significant amount of time is spent in class. Saving for college, even a small investment at a time can help make expenses more manageable.

1. Student A

Start saving age: _____2_____

Amount saved each month: _____$25_____

Amount saved each year: (a) $25 x 12 = $300

Total balance saved at age 18: (b) 16 x $300 = $4,800

2. Student B

Start saving age: _____18_____

Amount saved each month: _____$200_____

Amount saved each year: (a) _____

Total balance saved at age 18: (b) _____

3. Student C

Start saving age: _____7_____

Amount saved each month: _____$60_____

Amount saved each year: (a) _____

Total balance saved at age 18: (b) _____

4. Student D

Start saving age: _____10_____

Amount saved each month: _____$10_____

Amount saved each year: (a) _____

Total balance saved at age 18: (b) _____

5. Which student invested the highest amount each month?

6. Which student began saving at the youngest age?

7. Which student invested the least amount each month?

8. Which student had the lowest balance saved at age 18?

9. Was the answer to question 8 as you would expect?
   Why or why not?

EXPLORE

401K

Retirement Savings

Retirement is when a person no longer works a full-time job and has more time available during the day. Sounds relaxing, but working less hours requires a savings or investment account that can afford to pay the bills. This account is commonly named a 401K account. To better ensure that retirement is affordable people begin saving for retirement long before the planned retirement date. A surprisingly small amount can grow to a substantial amount of money.

Complete the missing information in the different retirement saving scenarios.

	Start Saving Age	Monthly Saved	Annual Saved	Years to Age 65	Total Amount Saved
1.	18	$24	$24x12 =$288	65-18 =47	$288x47 =$13,536
2.	30	$60	$		$
3.	21	$100	$		$
4.	25	$15	$		$
5.	40	$140	$		$
6.	28	$50	$		$
7.	22	$40	$		$
8.	45	$200	$		$
9.	27	$150	$		$
10.	20	$30	$		$

11. What conclusion can be made from the age a person begins to save and the total amount saved at retirement?

COMPUTE

www.YMBAgroup.com

Double Bar Charts

A double bar chart is a useful way to compare two different types of data. The data can be evaluated on its own, or as it relates to the second type of data being tracked on the chart. Below are two tables of data. Plot the data on the empty chart provided.

Gaston Gas Stations

	Gas Prices Per Gallon	Total Gas Customers (in hundreds)
Jan	$3.50	3.0
Feb	$3.75	3.5
Mar	$3.25	3.25
Apr	$3.50	3.0
May	$3.00	4.0
Jun	$3.50	3.25

Gaston Gas
Number of Customers Compared To Gas Price

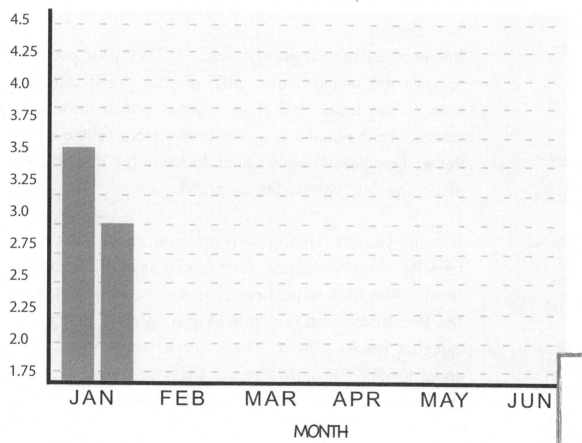

The Drawing Board

Pie Chart

A pie chart is useful for a quick glance at data to compare the amount of items to each other. The sections of a pie chart equal 100%. The larger sections of a pie chart show a more significant percentage of the overall total.

Read the clues and complete the pie chart.

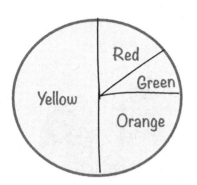

Imagine you have 100 jellybeans. 12 are green, 13 are red, 25 are orange, and the rest are yellow. Complete the pie chart to show the distribution of jellybean colors.

A meteorologist in Anytownville has researched the annual rainfall for last year. The findings show that 25% of the rain was in the first quarter of the year, 25% in the second quarter, 40% in the third quarter and 10% in the last quarter. Complete the pie chart to show the distribution of rain.

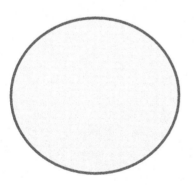

Imagine you have a garage sale. 35% of your profit comes from selling books. 50% of your profit comes from selling toys. 15% of your profit comes from selling movies. Your remaining profit comes from selling other items. Complete the pie chart to show the distribution of garage sale categories by profit.

Imagine you are talking with a friend about your favorite baseball teams. Your friend says his favorite team is the Blue Rays, half as much the friend likes the Red Birds. The friend likes the Patriots just as much as the Red Birds. These three teams equal 100%. Complete the pie chart to show the distribution of favorite baseball teams.

APPLY

www.YMBAgroup.com

Currency
Exchange Rates

Currency, also known as money, has a value that changes each day. Often the difference is a small amount, but in times of economic or government difficulties the changes may be larger. While in another country for vacation or work it is common to pay for goods and services using the currency (money) of that country. For example, a person from the USA when traveling to Mexico would exchange USA money (dollars) for Mexican money (pesos). A currency exchange is used to value the money of a country so it can be traded for money from another country.

USD = United States Dollar   CAD = Canadian Dollar

USD = 1.1870 CAD   CAD = .8424 USD

The exchange rate format has the first country shown equal to $1 of the country currency. The second number shown is the amount of the second country currency needed to trade for 1 dollar of the first country currency. For example, $1.19 Canadian Dollars would trade for $1 of United States currency.

What if a person using United States dollars wanted to buy Canadian currency?

The formula would be: 1 divided by CAD exchange rate

$$1 \div 1.1870 = .84 \text{ cents}$$

It would cost 84 cents in United States currency to trade for 1 Canadian Dollar.

Convert the following currency equations between United States and Canada.

1. USD = 1.2435 CAD
    How many Canadian Dollars to buy $1 of US currency? $ _____

2. USD = 1.1485 CAD
    How many Canadian Dollars to buy $1 of US currency? $ _____

3. USD = 1.033 CAD
    How many United States Dollars to buy $1 of Canadian currency? $ _____

4. USD = 1.2567 CAD
    How many United States Dollars to buy $1 of Canadian currency?
                                                          $ _____

COMPUTE

The Drawing Board

Currency Conversions

When traveling outside of your home country it is likely you would require money in the same currency as the local country. To trade your money for local currency, a currency conversion is necessary. Since the value of money from different countries varies, a currency conversion is used for trading money to obtain money of equal value.

US Dollar   USD $1.00

Australian Dollar	AUD	1.223
British Pound	GBP	.6482
Euro	EUR	.842
Japanese Yen	JPY	120.42
Swiss Franc	CHF	.996519

The chart shows the currency rate to purchase one United States Dollar. Complete the currency conversions below to compute how much of each currency in the chart would be purchased by one United States Dollar.

1. 1 AUD costs __.82__ cents in USD.

     1 ÷ 1.223 = .81767

          .81767 rounds to .82

2. 1 GBP costs _____ USD.

3. 1 EUR costs _____ USD.

4. 1 JPY costs _____ USD.

5. 1 CHF costs _____ USD.

6. 1 USD buys __$1.22__ AUD.

     chart shows AUD 1.223

     round to currency format = $1.22

7. 1 USD buys _____ BBP.

8. 1 USD buys _____ EUR.

9. 1 USD buys _____ JPY.

10. 1 USD buys _____ CHF.

COMPUTE

www.YMBAgroup.com

The Drawing Board

Number Challenge

Travel from the library to the beach. Select the path that will result in the highest number when you reach the beach. Add or multiply the along the path as you reach each number while traveling to the beach.

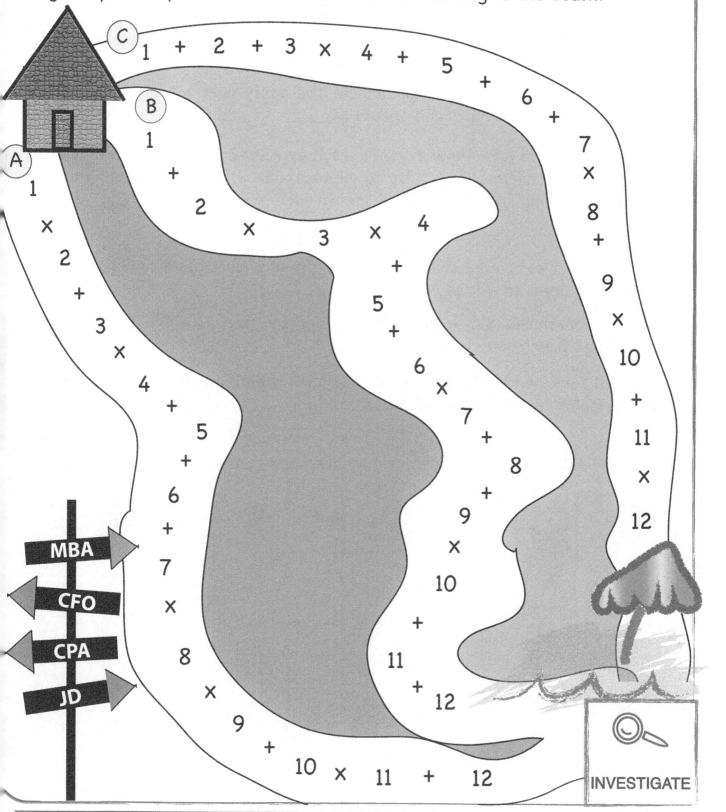

The Drawing Board

TOOLBOX

total   multiply   MBA   add   subtract   bar   pie

Down

1. A friend wants to track the average monthly temperature over the past year. The friend will best display the information in a _____ chart.

2. A total of 6 friends will visit for dinner. The party host will offer each friend three slices of pizza. The friend should _____ to find total slices needed.

5. A friend wants to know how many hours they read books this week. Day 1 they read for 20 minutes, day 2 for 40 minutes, day 3 for 30 minutes. The friend wants to find the _____ time spent reading.

Across

2. A friend just completed college and will attend a business school to seek their next degree, known as an _____ .

4. A friend purchases $10 in ice cream and pays with a $20 bill. The method to determine how much change is received is to _____.

6 A friend surveys 100 children at a waterpark regarding their favorite pets. The information is gathered and presented in a _____ chart.

7 A friend is shopping for party hats. The party will have 8 boys and 8 girls. The friend will _____ to determine how many party hats to purchase.

Notes

YMBA Business Math Review

Congratulations on completing the YMBA Business Math workbook.
Consider the questions below to demonstrate all you have learned.
Write your answers in the spaces provided on the answer page.

1. Which fraction is the greatest?

   (A) $\frac{1}{3}$ (B) $\frac{6}{8}$ (C) $\frac{3}{4}$ (D) $\frac{8}{10}$

2. Using Pemdas, solve: 21 - 9 x 2 + 12

   (A) 15   (B) 2   (C) 14   (D) 18

3. A monthly investment is made of $40, how much is saved in one year?

   (A) $480   (B) $400   (C) $40   (D) $1240

4. A total saved in three years was $3,600. How much was saved per month?

   (A) $360   (B) $100   (C) $300   (D) $120

5. Checking account balance: $765. What is the balance after writing a $120 check?

   (A) $585   (B) $620   (C) $565   (D) $645

6. When you deposit into a savings account the balance will:

   (A) Increase   (B) Decrease   (C) Stay The Same (D) Balance

7. Which of the following is not part of writing a check?

   (A) Date   (B) Birthday   (C) Memo   (D) Dollar Amount

8. Writing a check will cause the running balance to:

   (A) Increase   (B) Balance   (C) Decrease   (D) Stay The Same

9. Lemons are 40 cents a pound. You buy 32 ounces of lemons. The cost is:

   (A) $1.40   (B) .80 cents   (C) .32 cents   (D) $1.20

10. Bananas are $2.65 a dozen. You purchase 24 bananas. The cost is:

(A) $5.30   (B) $4.65   (C) $4.80   (D) $5.65

11. Milk costs $3.97 per gallon. A coupon is buy two get one free. The cost is:

(A) $6.94   (B) 11.91   (C) $7.37   (D) $7.94

12. A purchase totals $54.85 plus 6% sales tax. The total cost is:

(A) $58.14   (B) $56.85   (C) $60.85   (D) $57.59

13. An item costs $34.80 and with sales tax the cost is $36.24. Sales tax equals:

(A) $1.64   (B) $1.54   (C) $1.44   (D) $1.34

14. A recipe needs 25% of a dozen eggs. How many eggs are needed?

(A) 3   (B) 4   (C) 5   (D) 6

15. A $50,000 loan is given by a bank for one year plus 11% interest. Interest equals:

(A) $5,500   (B) $5,000   (C) $550   (D) $500

16. A savings account has $5,700. Equal investments for the past 60 months are:

(A) $85   (B) $95   (C) $65   (D) $75

17. How many quarters are in 5 years?

(A) 4   (B) 8   (C) 16   (D) 20

18. Rent payments of $1,585 per month paid for 18 months equal:

(A) $15,850   (B) $19,020   (C) $28,530   (D) $32,750

19. An invoice shows 18 books priced at $15 each. The total book cost is:

(A) $153   (B) $185   (C) $270   (D) $330

20. A credit card payment is due May 4. The payment made May 8 will have a:

(A) Credit   (B) Late Fee   (C) Debit   (D) Balance

21. A credit card balance is $3,543 with 8% interest. Interest is equal to:

(A) $28.43   (B) $244.83   (C) $283.44   (D) $354.52

22. The solution is 125. What is the correct exponent?

(A) 5^1 (B) 5^2 (C) 5^3 (D) 5^4

23. Company profit less dividends = $64,400. 10,000 shares outstanding. EPS is:

(A) $4.04   (B) $4.40   (C) $6.20   (D) $6.44

24. Solve: 4^3

(A) 12   (B) 24   (C) 48   (D) 64

25. Solve: $\sqrt{64}$

(A) 4   (B) 7   (C) 8   (D) 16

26. A factory produces 12 bottles, each 36 ounces, per hour. How many ounces tota

(A) 49 ounces   (B) 48 ounces   (C) 136 ounces   (D) 432 ounces

27. Each work shift is 8 hours. A factory runs 24 hours. How many shifts are there?

(A) 3   (B) 4   (C) 8   (D) 12

28. A wholesale cost is $2, retail price $3.65. 25 items sell, what is the profit?

(A) $36.50   (B) $41.25   (C) $73.00   (D) $91.25

29. A $17,500 truck is depreciated over 7 years. The yearly straight depreciation is:

(A) $2,000   (B) $2,300   (C) $2,500   (D) $2,750

30. Car A: $255.80 monthly for 5 years   Car B: $197 monthly for 7 years

(A) A costs more than B   (B) B costs more than A   (C) Equal amounts

31. Gas costs $62 month 1, $74 month 2, $68 month 3. The average is:

(A) 62   (B) 68   (C) 70   (D) 74

32. A truck travels 1152 miles at 32 miles per gallon. Total gallons used:

(A) 30   (B) 32   (C) 34   (D) 36

33. A 650 square foot room is having $7.32 per square foot carpet installed. The cost is:

(A) $4,758   (B) $5,265   (C) $6,500   (D) $7,320

34. A 7 x 9 foot garden is installing a fence around the perimeter. The total fence in feet:

(A) 32 feet   (B) 36 feet   (C) 63 feet   (D) 128 feet

35. The cost to fill a gas tank is $41.92 and the tank holds 16 gallons. The gas price is:

(A) $2.62   (B) $3.02   (C) $3.12   (D) $3.42

36. An hourly pay rate of $10.65 during a 38 hour workweek before taxes earns:

(A) $372.75   (B) $394.85   (C) $404.70   (D) $410.85

37. Which of the following is not a type of graph?

(A) Bar   (B) Double Bar   (C) Pie   (D) Square

38. Which is the more conservative low risk investment?

(A) Bonds   (B) Mutual Funds (C) Stocks   (D) All equal risk

39. Saving 12 years for college results in $24,480. The monthly savings amount was:

(A) $160   (B) $170   (C) $180   (D) $190

40. USD = 1.254 CAD   How much does it cost in USD to trade for 1 Canadian dollar?

(A) .75   (B) .80   (C) .95   (D) $1.24

This page intentionally left blank.

Y.M.B.A. Business Math Review Student Test Sheet

Consider the questions on the previous four pages.
Write your answers in the spaces provided below.

1. _____

2. _____

3. _____

4. _____

5. _____

6. _____

7. _____

8. _____

9. _____

10. _____

11. _____

12. _____

13. _____

14. _____

15. _____

16. _____

17. _____

18. _____

19. _____

20. _____

21. _____

22. _____

23. _____

24. _____

25. _____

26. _____

27. _____

28. _____

29. _____

30. _____

31. _____

32. _____

33. _____

34. _____

35. _____

36. _____

37. _____

38. _____

39. _____

40. _____

Certificate of Completion

Presented To

_____

Upon Successful Completion

of the

Youth Master of Business Administration

BUSINESS MATH

Presented By

_____

Date

_____

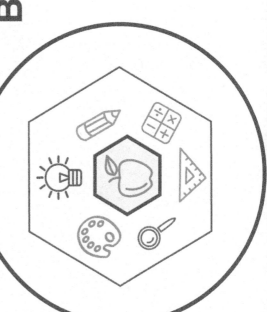

Y.M.B.A. Business Math Drawing Board Answer Key

Page 8: (1) 25% (2) 20% (3) 14.3% (4) 50% (5) 12.5% (6) 20% (7) 87.5% (8) 16.7% (9) 62.5% (10) 25% (11) 1/4 (12) 2/5 (13) 41/50 (14) 3/4 (15) 3/10 (16) 1/2 (17) 16/25 (18) 17/100 (19) 1/8

Page 9: (1) -2 (2) 23 (3) 28 (4) 8 (5) 22 (6) 60 (7) 39 (8) 54

Page 10: (1) $25, $75, $300 (2) $26.67, $80, $320 (3) $16.67, $50, $200 (4) 50, $150, $600 (5) $36.67, $110, $440 (6) $13.33, $40, $160 (7) $45, $135, $540 (8) $20, $60, $240 (9) $100, $300, $1,200 (10) $10, $30, $120 (11) $166.67, $500, $2,000 (12) $18.33, $55, $220 (13) $70.

Page 11: (1) $1,000 (2) $875 (3) $820.66 (4) $702.26 (5) $452.26 (6) $383.54 (7) $300.64 (8) $88.64 (9) $1,476.72

Page 12:

					624 00
1112	8/2	American Tennis Club	45 50		578 50
--	8/6	Weekly Chore Allowance		5 00	583 50
1113	8/9	Doctor Visit	15 00		568 50
1114	8/12	Car Payment	185 00		383 50
1115	8/15	School Books	8 75		374 75
1116	8/16	Repay loan from friend	60 00		314 75
	8/18	Sold Comic Books		5 00	319 75
	8/21	Washed Car		5 00	324 75
1117	8/23	Grocery Food	21 38		303 37
	8/27	Coins Deposit		7 45	310 82
	8/29	Receive Birthday Check		30 00	340 82
	8/31	Monthly Interest Deposit		.82	341 64

Page 13:

Page 14: (1) $8.96 (2) $9.50 (3) $27.80 (4) Customer 3 spent the most.

Page 15: (1) $22.35

Page 16: (1) $60 (2) $20 (3) $40 (4) $16 (5) $0 (6) $136 (7) $136x.07= $8.84 (8) $144.84 (9) $160 (10) $15.16

Page 17: (1) $3.87 (2) $122.86, $27.14 (3) $90.10, $9.90 (4) $275.18, $24.82 (5) $33.50, $6.50 (6) $70.47, $9.53 (7) $132.06, $7.94

Page 18: (1)2(2)4(3)6(4)1(5)4(6)12(7)2(8)16,32(9)5(10)10(11)15(12)2 1/2 (13)10(14)30(15)5(16)40,80

Page 19:(1)$5,520(2)$16,560(3)$13,750(4)$12,880(5)$21,390(6)$22,000

Page 20: (1) $85.71, $600, $28,800 (2) $32.14, $225, $10,800 (3) $42.86, $300, $14,400 (4) $62.50, $437.50, $21,000 (5) $68.86, $482, $23,136 (6) $51.25, $358.75, $17,220 (7) $57.14 $400, $19,200 (8) $23.21, $162.50, $7,800 (9) $28.93, $202.50, $9,720 (10) $50.36, $352.50, $16,920 (11) $71.43, $500, $24,000 (12) $41.07, $287.50, $13,800 (13) answers vary.

Page 21: (1)$1,112 (2)$1,242

Page 22: (1) 5369,3/17/14(2) party rentals(3)Orlando, FL(4)18 (5) $5.95(6)25(7)$1.15(8)$9.43(9)$47.64(10)white

Page 23: (1) Georgia (2) $25 (3) $150 (4) partyhatsofun.com (5) doggie bath and fresh food grocery (6) $16.50 (7) store return (8) a dog (9) $73.82

Page 24: (1) $318.48 (2) $426.24 (3) $303.13 (4) $612.71 (5) $25.60 (6) number 3 (7) number 4 (8) highest balance had a lower interest rate.

Page 25: (1) 32 (2) 81 (3) 25 (4) 512 (5) 2,187 (6) 1,024 (7) 1 (8) 9 (9) 343 (10) 0 (11) 243 (12) 36 (13) 64 (14) 121(15) 343 (16) 216

Page 26: (1) 8x8 (2) 11x11 (3) 3x3 (4) 7x7 (5) 12x12 (6) 6x6 (7) 9x9 (8) 2x2 (9) 20x20 (10) 15x15 (11) 25 (12) 1 (13) 100 (14) 49 (15) 625 (16) 64 (17) 144 (18) 6.25

Page 27: (1) 12 (2) 120 (3) 19.2, 192 (4) 24 (5) 15 (6) $11.20 (7) $7.00 (8) $12.80, $8 (9) 8 ounces

Page 28: (1) 3 minutes (2) 3 (3) 5 (4) 20 (5) 160

Page 29:(4)$5(3)$245(7)$245(8)$170(9)$15(14)$1,638(15)$1,638(17)$588(18)$3.40(23)$10,874.80(24)$10874.80(26)$2,714.80

Page 30: (2)$2,000(3)$700(4)$2,500(5)$3,150(6)$28,067

Page 31: (1)12(2)36(3)60(4)84(6)$15,120,$22,440(7)$13,860,$10,728(8)$18,480,$21,000(9)$18,432,$32,088 (10) varies (11)what can afford to pay

Page 32: (1) $316 (2) $948 (3) $4,056 (4) 22.57 hours

Page 33: (2)36,$108(3)31,$124(4)$900(5)$935(6)$1,413.60(7)$1422.40(8)$1,669.36(9)$1,984(10)$1,350

Page 34:(2)$150(3)$288(4)$310(5)$384(6)$168(7)$256(8)12,$36(9)16,$32(10)18,$72(11)28,$56

Page 35:(1)114,4.56,$16.64(2)185,7.4,$27.01(3)149,$5.96,$21.7 (4)175,7,$25.55

Page 36: (1) $54.60 (2) 30x15=450 miles (3) $63.36 (4) 450 miles (5) 300 divided by 25 =12 (6) $38.40 (7) 18.75 mpg, 20 mpg (8) car F with 319 miles (9) car H with 432 miles.

Page 37:(2)$1,354.17,$1,069.79(3)$520.83,$411.46(4)$729.17,$576.04(5)$1,708.33,$1,349.58(6)$2,958.33, $2,337.08(7)$375,$296.25(8)$401.88, $333.29(9)$1,331.88,$1,052.19 (10) $755,$596.45

Page 38:(2) $83.16,$312.84 (3) $67.83, $255.17 (4) $68.21,$256.59 (5) $34.02, $127.98 (6) $73.29,$275.71 (7) $105.46, $396.74 (8) $164.34,$618.21 (9) $221.93, $834.87 (10) $70.42, $264.93 (11) $53.93, $202.87 (12) $87.83,$330.42

Page 39: (B) $48.40, $30.80, $13.20 (C) $90.75, $57.75, $24.75 (D) $68.20, $43.40, $18.60 (last) $251.35

Page 40: (1) stock price is going up, increasing

Page 41: p/e ratios: 22, 11.84, 18.94,17.31 (1) company a, higher p/e ratio.(2) company B, lower p/e ratio (3) A (4) B

Page 42: (2) $200+(($200x.05)x(25-21)) =$200+($10x4)=$200+$40 = $240 (3) $50 +(($50x.04)x(40-25))=$50+($2x15)=$50+ $30=$80 (4) $500+(($500x.02)x(28-11)= $500+($10x17)=$500+$170=$670

Page 43: (1)medium (2)high (3)low (4)high (5)medium

Page 44: (1)$21.13 (2) $100 (3)$65.66 (4) high shows more paid to investors

Page 45: (2) (a) $2,400 (b) $2,400 (3) (a)$720 (b) $7,920 (4) (a)$120 (b) $960 (5) b (6) a (7) d (8) d (9) answers vary.

Page 46: (2)$720, $35, $25,200 (3)$1,200 $44, $52,800 (4) $180, $40, $7,200 (5) $1,680, $25, $42,000 (6) $600, $37, $22,200 (7) $480, $43, $20,640 (8)$2,400 $20, $48,000 (9) $1,800, $38, $68,400 (10) $360, $45, $16,200 (11) start younger

Page 47

Page 48:

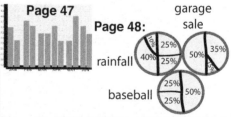

garage sale, rainfall, baseball

Page 49: (1)$1.24(2)$1.15(3)$.97(4)$.80

Page 50: (2)1.54(3)1.54(4)1.19(5)1(6)1.22(7).65(8).84(9)120.42(10)1

Page 51: (a)30,218 (b)3483 (c)41532

Page 52: (1) bar (2) MBA (3) multiply (4) subtract (5) total (6) pie (7) add

www.YMBAgroup.com

1. D

2. A

3. A

4. B

5. D

6. A

7. B

8. C

9. B

10. A

11. D

12. A

13. C

14. A

15. A

16. B

17. D

18. C

19. C

20. B

21. C

22. C

23. D

24. D

25. C

26. D

27. A

28. B

29. C

30. B

31. B

32. D

33. A

34. A

35. A

36. C

37. D

38. A

39. B

40. B

Benefit from 100 top tips and tricks that will enhance the effectiveness and enjoyment of your Internet-based virtual classroom presented by virtual teacher, recruiter and trainer, L.J. Keller. A must-have book for anyone considering, or currently teaching, English virtually as a second language! Benefit from ideas, techniques and examples for English lanugage learners. These easy to implement concepts can enhance your classroom and effectively increase your students comprehension. Students also enjoy learning in this active learner classroom environment. Ideas are presented with clarity using examples that can provide you with a competitive advantage in the virtual classroom. Enjoy this all-in-one solution to help you launch and sustain amazing student results. **Are you ready to be an amazing virtual teacher?**

Ready to define your S.P.A.C.E.?
Do you know your 4 E's?

www.YMBAgroup.com

ISBN 9781690614036

9 781690 614036

Do you know someone who would like to work from home?
Virtual teaching is a wonderful option!
Work from home.
Flexible schedules.
Amazing students!

This one book can help you quickly achieve a successful virtual classroom.

100 Tips are ready to assist!

Recruiting for virtual teachers with a four year college degree, any major at:
www.YMBAgroup.com

100 Virtual ESL Teacher Tips and Tricks

Proven Ideas For Student Enjoyment and Success!

L.J. Keller, M.B.A.

SPLASH: A waterpark mystery

Moore Mysteries | Book 1

The champion is part of a secret plan! Solve the mystery as you meet Battle, twins Rachel and Reese, Zack and Morgan as they travel the United States with their parents. In this first book in the series the family begins a road trip adventure. The first *family fun stop* finds a mystery the family works together to solve. Join the family as they race to tell the judges. What was the secret plan? How will the kids find the judges to stop the results in time? What is discovered?
Grades 2-4/Ages 6-8/Early Chapter Book

Look Inside!

Skill Builder practice and
a Book Quiz Included

**Engaging Reading Books
plus
Skill Builders & a Book Quiz
An easy way to demonstrate
learning accomplishments.**

57 Days
Miriam Short Sails with William Penn

57 Days Miriam Short Sails With William Penn

Learning Book Quiz Included!

YMBA Days In History Series

Join Miriam, Adam and Anne Mary, on their journey from England to America with William Penn to see the land he was granted in the New World by the King of England. Exciting history based on actual people and events. Experience the triumphs, struggles, loss and dreams while traveling across the Atlantic Ocean to a new home. Discover the path so many experienced as they left their home for America. Details vividly paint a picture of the conditions on the ship and the difficult days along the way. What challenges did they endure?
What were the fears and hopes of the young adults?
An exciting historical adventure of the journey to America.
Join Miriam on her voyage with her family and William Penn.
GRADES 6-10/AGES 11-15/ FACTION CHAPTER BOOK

Y.M.B.A.
Single
Topic
Learning
Workbooks

Lesson Pages,
Worksheets,
A Quiz
and
A Certificate

Learn
Life Skills
&
Business
with
Y.M.B.A.

Made in the USA
Monee, IL
03 August 2020

37477796R00037